NASA's Ships and Planes

by Patrick H. Stakem

2017

Number 12 in the Space series.

Table of Contents

Introduction..3
Author...3
NASA's History..5
   Kennedy Space Center..6
NASA's Airplanes...7
   X-15...7
   STS-Shuttle..9
   NASA's Calibration aircraft..11
Flying Telescopes..13
   Goddard...14
      STADAN..14
      NASCOM...15
   Support from orbit...16
      TDRSS..17
The NASA Navy...17
   Tracking and Instrumentation Ships..17
   Recovery Ships...21
   Shuttle External Tank transport...24
Airborne Observatory...25
Wrap-up..26
Bibliography..27
Resources...32
Directory of U. S. Navy ships used to recover NASA crewed missions. 34
Glossary of Aerospace Terms..48
Glossary of Nautical Terms and Phrases..55

Introduction

This book discusses the resources and infrastructure that NASA developed and applied to support space missions in the early 1960's. When the first U. S. satellite, Vanguard, went up in 1958, there was no world-wide network of tracking stations. Thus, a series of ground stations, tracking ships, and tracking aircraft were required. When the crewed capsules of the Mercury, Gemini, and Apollo programs splashed down in the ocean, they were retrieved by U. S. Naval vessels for NASA. NASA has a couple of Ocean-going tugs to retrieve the solid rocket boosters used in the Shuttle Program. In addition, the size of the Shuttle external tank required water transportation on a barge, from the assembly site to the launch site. The Shuttle itself was flown from place to place on top of a specially modified 747 aircraft.

We will discuss the ships and planes that supported NASA missions, and see how these evolved over the years.

Author

Mr. Patrick H. Stakem has been fascinated by the space program since the Vanguard launches in 1957. He received a Bachelors degree in Electrical Engineering from Carnegie-Mellon University, and Masters Degrees in Physics and Compute Science from the Johns Hopkins University. At Carnegie, he worked with a group of undergraduate students to re-assemble, modify, and operate a surplus missile guidance computer, which was later donated to the Smithsonian. He was brought up in the mainframe era, and was taught to never trust a computer you could lift.

He began his career in Aerospace with Fairchild Industries on the ATS-6 (Applications Technology Satellite-6) program, a communication satellite that developed much of the technology for the TDRSS (Tracking and Data Relay Satellite System). He followed the ATS-6 Program through its operational phase, and worked on other projects at NASA's Goddard Space Flight Center including the Hubble Space Telescope, the International Ultraviolet Explorer (IUE), the Solar Maximum Mission (SMM), some of the Landsat missions, and Shuttle. He was posted to NASA's Jet Propulsion Laboratory for Mars-Jupiter-Saturn (MJS-77), which later became the *Voyager* mission, and is still operating and returning data from outside the solar system at this writing. He initiated and lead the international Flight Linux Project for NASA's Earth Sciences Technology Office. He is the recipient of the Shuttle Program Manager's Commendation Award, and has completed 42 NASA Certification courses. He has two NASA Group Achievement Awards, and the Apollo-Soyuz Test Program Award.

Mr. Stakem has been affiliated with the Whiting School of Engineering of the Johns Hopkins University since 2007, and Capitol Technology University. Mr. Stakem supported the Summer Engineering Bootcamp Projects at Goddard Space Flight Center for 2 years.

NASA's History

NASA was formed to respond to the Soviet Union's launch of the first artificial satellite of the Earth. President Kennedy focused America's efforts on a race to the Moon, then seen as the goal for Mankind. The Space Race was off and running, with seemingly unlimited resources devoted to that goal. Kennedy stated the national goal of landing on the moon in the 1960's. This goal was achieved, but unfortunately Kennedy was not alive to see it. NASA partners with the National Oceanographic and Atmospheric Administration for weather, climate, and disaster monitoring from space. It is also the focal point for exploration of the rest of the Universe, the other planets in our solar system, stars, galaxy's and whatever is out there. NASA co-operates with other space-faring nations on joint projects.

NASA continued it's leadership in aeronautics, and built the infrastructure we have come to rely on, the weather satellites, and the communication relay satellites, (a concept of Science Fiction Writer, Sir Arthur C. Clarke,) and the Global Positioning System (GPS), so we don't need hard-to-fold paper maps any more.

The National Aeronautics and Space Administration supports aerodynamic and space flight for non-military purposes. In their own words, NASA's Vision is 'We reach for new heights and reveal the unknown for the benefit of humankind.' That's what they have been doing for nearly 60 years now, since President Eisenhower established the agency in 1958. It was an expansion of the National Advisory Committee on Aeronautics (NACA), which at the time was 40 years old. NASA is still involved in winged flight, but is best know for its work in Space. Even as I type, Astronauts aboard the International Space Station are working to improve our life on Earth, and our understanding of our place in the Solar System and the Galaxy. NASA manned missions have been to the Moon, and robot spacecraft have visited all the planets of our solar system. There's a lot more work to do.

We'll stretch this a little, and mention NASA's helicopter on Mars. It's name is Ingenuity. It has a co-axial rotor. It is parto f the Mars 2021 mission, and was carried by the Perseverance lander. It first fles on April 19, 2021. It still operational, after 525 days after its first flight. It was build at JPL.

Kennedy Space Center

Kennedy Space Center, in Florida, is NASA's major spacecraft launch facility. It hosts uncrewed launches, and the iconic Apollo and Shuttle missions. Vehicles from Kennedy launch to the south east. This allows spent lower stages to be jettisoned into the ocean. The Shuttle's rocket solid rocket boosters were recovered by ships from the Space Center's port, and returned for refurbishing and reuse. As of this writing, crewed spacecraft are scheduled to lift off from Kennedy launch pads again, as the Orion capsule and Space Launch System booster are getting ready for first flight.

NASA's Airplanes

Since the first "A" in NASA is aeronautics, we expect NASA to have experimental and test aircraft. Most aeronautic research takes place at the Armstrong Flight Research Center on the site of Edwards Ai5 Force Base. This is where the X-15 research was based. The NASA Langely Research Center in Hampton, Virginia was the first NASA Center, established in 1917, and focused on flight. Actually, back then the agency was NACA – the National Advisory Committee for Aeronautics. Back then, only a few crackpots such as Robert Goddard, Werner von Braun, and Willy Ley were interested in rockets.

The development and maturation of the rocket airplane would lead to two milestones in the post-war era. The first was the breaking of the sound barrier by Chuck Yeager in 1947. Going trans-sonic removed a barrier to speed. The next limit to be broken was the ability to get to 100 km in altitude, beyond which was space. This was accomplished by the X-15 project.

X-15

The X-15 was a manned hypersonic rocket-powered aircraft, designed to fly to the edge of space. You could earn your astronaut badge in an X-15. The international standard is 100 km or 62 miles. The USAF uses the more liberal 50 miles. There were 13 flights greater than 50 miles, and two greater than 62 miles. Both flights to "space" were made by Joseph A. Walker in 1963. The X-15 was carried to altitude under the wing on a B-52

Mothership. After it was released, its rocket engine was ignited, pushing it up out of most of the atmosphere. There were no air-breathing engines. It did a "dead-stick" landing. There were two flight to the 100 km altitude, both by Joseph A. Walker in 1963. The X-15 flights were USAF, U.S. Navy, and NASA sponsored.

The X-15 holds the official world's record to the highest speed recorded by a manned, powered aircraft 4,250 mph (Mach 6.7) at 102,100 feet altitude. That was in 1967. It is officially the world's first space plane.

The X-15 was developed from a concept by Walter Dornbeger for NACA in 1954. He was one of the captured German scientists that formed the core of the rocket team. Dornberger served as the military officer in charge of the German rocket program, and he came to the U.S. with von Braun.

The X-15 operated in two distinct domains, and had both aerodynamic control surfaces, and rocket thrusters. The plane included a pilot ejection seat, usable up to Mach 4, and 120,000 feet. The main engines were dual *Reaction Motors* XLR11 units, using alcohol and liquid oxygen to achieve a total of 16,000 pounds of thrust. Earlier, a single XLR11 pushed the Bell X-1 to be the first aircraft to exceed the speed of sound (Mach1). Later, the X-15 was fitted with the upgraded XLR99 for 57,000 pounds of thrust. Over 175 flights were made in that configuration. Three X-15's made a total of 199 test flights, the last in 1968. Twelve pilots flew the planes, including future astronaut Neil Armstrong, who would go on to become the first man on the moon.

STS-Shuttle

The Shuttle, or Space Transportation System, flew to space like a rocket, and landed like an airplane. Actually, like a glider. Early Shuttle studies considered air breathing engines for more control upon reentry, but these were not implemented due to weight issues. After re-entry, the Shuttle could glide to a runway at Kennedy Space Center, but had very limited cross-range capability. Another option was to use the Dryden flight facilities vast expanses of hard desert. In that case, the Shuttle was brought back to the Kennedy Space Center on the back of a specially modified 747 carrier aircraft.

The Shuttle was similar in size to a DC-9 jetliner, and was of a lifting-body, going back to the German *Silbervogel* design of World War-II. The entire lower surface of the craft acted as a wing.

At launch, the STS consisted of the winged Shuttle vehicle, a large liquid fuel and oxidizer external tank, and two solid rocket boosters. The solid rocket casings were retrieved from the ocean, refurbished and reused. The external tanks were not recovered, and were targeted away from shipping lanes in the Pacific and Indian oceans.

There was a mock-up, a prototype, and five flight units. Two of the flight units were destroyed, one at launch, one at reentry, both with loss of crew.

The Shuttle Orbiter rode the side of a large fuel and oxidizer tank to orbit, assisted by fall-away solid boosters to get everything going. The Shuttle Orbiter had three engines, fed from the large external tank. When the engines had burned sufficiently to achieve orbit, the Orbiter separated from the tank. The Orbiter continued to its destination altitude. The engines went with it, but no longer had a source of fuel or oxidizer. The Orbiter could adjust its orbit somewhat with its OMS (orbital maneuvering system) engines, using fuel onboard. There were also (reaction control system) RCS engines to adjust attitude. Upon reentry, the Shuttle flew in a nose-up attitude, as the bottom of the craft and wings were covered in heat-resistant tiles. After sufficient atmosphere was reached, the Shuttle's aerodynamic control surfaces could be used, and the Orbiter was flown like a plane to a runway landing. Well, like a 165,000 lb glider.

One of the Shuttle Carrier aircraft, a specially adapted Boeing 747, can be seen at Palmdale (CA)'s Joe Davis Heritage Airpark. NASA retains ownership of the aircraft. The other carrier aircraft was placed at Space Center Houston, with a Shuttle mockup on top.

What we will discuss next are the fleet of tracking aircraft that that "filled in the gaps" for telemetry coverage.

Not belonging to NASA, but leased by them was the AeroSpacelines Supper Guppy. This somewhat bloated craft would carry the Saturn-IVB stage. It was based on the Boeing 377. It weas lengthened to 141 feet, and the craft had an inside diameter of 25 feet. It could carry 54,000 pounds at 300 miles per hour. It was later modified with newer turboprop engines. There were two units in service. The turbine powered unit had a larger cargo section as well.

Where are they today?

Super Guppy N940NS is on display at the Pima Air & Space Museum at Davis–Monthan Air Force Base, Tucson, Arizona, US.

Super Guppy Turbine F-BTGV serial number 0001, was broken up in December 2020, with only the cockpit saved by the South Wales Aviation Museum.

Super Guppy Turbine F-BPPA is on static display at the Musée Aeronautique Aeroscopia, at the Airbus facility in Toulouse–Blagnac Airport, France.

Super Guppy Turbine F-GDSG, is on static display at the Airbus facility in Hamburg Finkenwerder Airport, Germany.

Super Guppy Turbine N941NA is still in service with NASA as a transport aircraft and is based at the El Paso International Airport, Texas, US.

NASA's Calibration aircraft

NASA's tracking stations used Calibration aircraft to check their systems. The planes carried the same transponder the spacecraft would. This allowed a checkout of the radio beacon, and tracking operation. The planes included a DC-3, a pair of DC-4's, and a Lockheed L-1649A Super Constellation. These were all equipped with the Mercury TT&C (tracking, telemetry, and command) equipment. The aircraft also participated in acceptance testing for antennas, and for calibration of the Minitrack stations. Working with the tracking ships, the calibration aircraft were used for the Dynamic Operations Test (DOT).

I have to pause here to tell you a story that will show you the professionalism and dedication of those tracking station crews. The author went to the Rosman Tracking station in North Carolina, from Goddard, to check on compatibility with the ATS-6 spacecraft. It was supposed to carry video from the Apollo-Soyuz Mission, the first time it had ever been done from orbit. When I got back to the control center at Goddard, they were testing the video reception.

As a calibration test article, the station had a plastic basket of fruit, that they pointed a camera at, and sent the image to Goddard. In the basement of Building 14, all the landlines from the various tracking stations converged, at NASCOM.

The NASCOM tech could not get the colors to come out correctly. After a few hours, the Rosman team admitted to spray-painting the plastic banana blue. A few rare swear words went over NASA voice communications that day. links that day.

Flying Telescopes

To get above the atmosphere, it is not required to have a satellite. High altitude aircraft and balloon-borne telescopes is a partial solution.

The first balloon borne astronomy mission was Stratoscope I in 1957. The Boomerang (Balloon Observations Of Millimetric Extragalactic Radiation And Geophysics) project looked at the cosmic background microwave radiation. It flew above 125,000 feet. First flight was in 1997. In 1998 and 2003, it ascended from McMurdo Station in Antarctica. The mission duration on those flights was 2 weeks. Another approach that has been used is to have a balloon carry a sounding rocket to altitude. These can be launched from anywhere, on land or water, and are still in use

.

The High Energy Focusing Telescope was a balloon payload looking for hard x-ray sources, in 2005. The High-resolution gamma-ray and hard X-ray spectrometer (HIREGS). flew from McMurdo Station, in Antarctica in 1991 and 1992.

NASA's Galileo Airborne Observatory was hosted on a Convair 990 aircraft. It was first used in 1965, but was unfortunately destroyed n a mid-air collision in 1973.

NASA's Kuiper Airborne Observatory used the larger Lockheed C-141 Starlifter. It operated from 1974 to 1995.

The Joint NASA-DLR Stratospheric Observatory for Infrared Astronomy (SOFIA) has been in use since 2010. It uses a Boeing 747SP with a 2.7 meter telescope. The aircraft is based at the Armstrong Flight Research Center in Pasadena, CA. The telescope is specially mounted to isolate it from movements of the plane. Missions are flown at 42,000 feet, which allows observation of Infrared above most of the atmosphere's blocking water vapor.

Goddard

The Goddard Space Flight Center (GSFC) is the hub of the NASA world wide communications Network, and the Lead Center for unmanned spaceflight. Located in Greenbelt, MD, it was dedicated in 1959 by rocket pioneer Dr. Robert Goddard's widow. Goddard is the lead NASA center for unmanned spacecraft. It has worked on hundreds of spacecraft projects, including the Hubble Space Telescope, and the upcoming James Web Space Telescope. Goddard has a nice Visitor's Center, open to the public.

STADAN

The spacecraft Tracking and data acquisition network was a NASA project in the 1960's. It was operated out of the Network Operations Control Center, in the basement of Building 14, at Goddard Space Flight Center. Early on, it could provide coverage for 15 minutes out of each 90 minute orbit. It eventually employed over 6,ooo people, with 24 locations on 5 continents. It was used to track

Sputnik, the first artificial satellite. Stadan was mostly obsolete by the 1980's

NASCOM

The NASA Communications System is housed in the basement of Building 14, on the GSFC Campus. It is the nerve center for NASA's World-wide communications links, using fiber optic landlines and satellite links. Before NASCOM, was the Minitrack system, starting with tracking Sputnik and Vanguard. It was the "minimum trackeable (object) network. The communications terminals were teletypes, and the data rate was 30 bits per second. For the Vanguard mission, several stations around the globe received telemetry, recorded this to analog magnetic tape, and mailed this to Goddard. There was something of a large time-lag involved. GSFC stored these tapes in a large warehouse. Backup tapes were kept at the stations. They're mostly unreadable now, due to deterioration over time.

By 1983, NASCOM was comprised of 139 stations world-wide, with 630 dedicated circuits connecting them with Goddard, using leased lines. NASA followed the emerging Internet model, having 5 world-spanning wide area networks, one of which was NASCOM. This was a packet-switched network, supporting NASA's legacy 1200 bit and 4800 bit datablocks. NASCOM data blocks were replaced by CCSDS packets of the Space Link Extension Service. Legacy NASCOM blocks are still supported, encapsulated in IP packets.

Minitrack did not support sending commands to the spacecraft. NASA starting developing a new global command and telemetry network, using lessons-learned from NORAD and the Air Force SAGE system. The Nascom project was started in 1960, and was beginning to take shape by 1964. It supported spacecraft tracking, data acquisition, commanding for near-Earth orbit. It incorporated JPL's Deep Space Network, and JSC's Manned Spaceflight Network.

Support from orbit

Satellite tracking and uplink/downlink always suffered from poor coverage, with a limited number of ground stations located at strategic points around the Earth, augmented by tracking ships. NASA contracted with Comsat Corporation to provide three communication satellites in synchronous orbit to provide links to the tracking ships for the Apollo missions.

A much better system evolved from a set of communications relay satellites up at geosynchronous altitude. It only required three satellites to give continuous coverage. They did keep an on-orbit spare. The proof-of-concept for this system was the early ATS-6 spacecraft. The author working on this project for Fairchild Industries who built it, and various contractors who flew it. The ATS-6 success in tracking satellites at lower altitudes led to the TDRSS System, and demise of most of the tracking ships.

TDRSS

Following on the success of the ATS-6 Mission, a systems of tracking an data relay satellites were placed into orbit, giving almost 100% coverage of satellites in low Earth orbit. TDRSS replaced the earlier NASA ground stations around the globe, and used a single ground station with three terminals at White Sands, New Mexico, relaying data to Goddard. There is a backup ground station on Guam, which is remotely controlled. Theoretically, you only need three spacecraft for complete coverage. Due to on-orbit spares requirements, and usage, there are nine satellites in orbit. The TDRSS were used for Shuttle missions as well, and currently supporst the International Space Station.

The NASA Navy

This section discusses the various naval vessels NASA used to support its missions. These were mostly needed to fill in gaps in radio coverage on the oceans.

Tracking and Instrumentation Ships

Since 2/3's of the Earth is covered in water, we certainly need tracking ships to allow for complete orbital coverage of satellites. These were called missile range instrumentation ships. Originally Navy World War-II oilers or Victory ships, the Navy had 23 of these from the 1950's through 2014. There are currently only two active in the fleet, the USS Invincible, and the USS Howard O. Lorenzen. The Soviet Navy had 8, of which one is active. The French had several, as do the Chinese. Even the USAF had a fleet of Missile Range Instrumentation ships.

A typical example is the USNS Vanguard (T-AGM-19), in service for missile tracking from 1964-1999. This ship was used to support the Apollo-Soyuz and the Skylab programs. The ship had large onboard tracking and telemetry antennas, which tended to make her somewhat top heavy. She is being "recycled" at a shipyard in Texas.

The ships served as nodes on the NASA Manned Space Flight Network, supporting the Mercury, Gemini, and Apollo missions. One ship filled a critical gap between the tracking stations at Bermuda and Antigua. This was critical, as the timing for the orbit insertion burn depended on actual time of launch, and launch vehicle performance. Another critical point was the departure from Earth orbit on the trans-lunar trajectory. This could take place anywhere in orbit, and the Atlantic, Pacific, and Indian oceans had to be covered. Apollo required up to 30-days on-station, to cover the return of Apollo to Earth orbit, and reentry. Two ships were assigned reentry duties in the landing area. NASA personnel on the ships were supported by the Military Sea Transport Service personnel, and DOD contractors.

At the peak of the Apollo era, there were three 19-Class ships, and two 6-class. The 19-class ships included the

USNS Vanguard, Redstone, and Mercury. These had 17 officers, 71 crew, and 122 technical personnel onboard. The 6-class ships were the USNS Watertown, and Huntsville. These carried 14 officers, 56 crew, and 72 technical personnel.

The ships were converted from World War-II Victory ships, or tankers. Large antennas were added to the deck for C-band radar and S-band tracking and data. An enhanced navigation system was added. In addition, a roll stabilization system was added to permit the steerable antennas to continuously track their target.

The 19-class ships were modified by removing the bow and stern of a tanker, and adding a new mid section for the electronics and crew. They were 595 feet long with a beam of 75 feet. Fully loaded, they displaced over 23,000 tons. They could achieve 17 knots, but tracking was usually done at 13 knots. They had an endurance capability of over 20,000 nautical miles.

The 6-class ships were derived from World War-2 era Liberty ships. They were 455 feet long, with a beam of 62 feet. They displaced 12,200 tons. They could achieve 16.5 knots. Their endurance was 10,000 nautical miles.

For the mission support personnel, amenities were provided, so they could focus on their jobs. Stewart's made the beds, and cleaned their quarters. There was a self-service laundry on board, a commissary, hospital, library, and private mess room (dining area, for you landlubbers).

The ship's Command Control System (CCS) was used to transmit commands to the spacecraft, and receive telemetry. This was only implemented on the 19-class. The ships also needed an accurate time reference, which was provided by a Rubidium standard, accurate to 5 parts in 10^{11} per year. The system was calibrated against the WWV System at the Bureau of Standards in Colorado (This was prior to the GPS system).

The 19-class ships carried a full mission control center, including a Command Communicator, four Vehicle Monitors (Flight Controllers) , an aeromedical monitor, and Flight Dynamics Monitor. This was a backup to the control center at Goddard, which was the backup to Houston.

The complement of computers onboard included Sperry-Univac 624B's and Univac 1218's. The 1218 was a militarized 418, in a cabinet 6 feet tall and weighing 775 pounds. It used 18-bit core memory, and could be configured with 4k to 64k of memory. In Navy use, it was a fire control computer, circa 1963.

The position and attitude of the ship were provided navigation and position information. This used the LORAN-C navigation system, gyro compasses, a navigation satellite receiver (probably the Navy's Transit), feeding the Integrated Navigation System. This later system included a star tracker. After a ocean floor survey, the ships deployed transponders on the ocean floor, which were used to define the ship's surface position.

After each mission, all telemetry data captured on magnetic tape were backed up, and put in secure storage, for return to GSFC.

For exploration of some of the "Ocean Worlds," NASA is considering a submarine that is deployed by a lander, after a hole is drilled in the ice. One candidate is Saturn's Moon Enceladus. These is a proposed mission called Enceladus Life Finder. Another is Jupiter Icy Moons Explorer, (JUICE).

Recovery Ships

NASA did not have specific ship resources to handle capsule recovery in an ocean landing. It received assistance from the U. U. Navy, who would pre-position an aircraft carrier and support ships in the target area. After the spacecraft splashed down in the water, a helicopter would be launched to retrieve it, and place it on the carrier's deck.

When the Space Shuttle was introduced, it was designed with two re-usable solid rocket boosters, that would drop into the ocean and be recovered by ships. They would then be returned to Cap Kennedy for refurbishment. The ships were the *MV Liberty Star* and *MV Freedom Star*. Each ship handled one booster. There was a 7,500 pound capacity deck crane to lift the end of the booster onboard.

These vessels were also employed to tow the barge with the Shuttle's external fuel tank from the assembly plant at Michoud, LA, to the Cape.

The Michoud facility is administratively a part of the Marshall Space Flight Center, and was built to support construction and integration of the massive Saturn-V moon rockets. These were so large they were delivered by barge to the Kennedy Space Center. For the Space Shuttle Program, Michoud was used to assemble the large external fuel tank. It is located east of New Orleans, Louisiana. It was established in 1961, and was originally a land grant to a local merchant from the King of France. It is named after Antoine Michoud, a Napoleon-era administrator in the area. The site was used as a sugar cane processing plant. Later, the site was used as a shipyard to construct Liberty Ships and Cargo aircraft during World War-II. During Hurricane Katrina, 37 employee's stay on site to monitor and protect the facility. These were awarded the NASA Exceptional Bravery Medal.

Covering more than 830 acres, it has one of the world's largest production buildings, the vertical assembly building, to support the assembly of the Shuttle external fuel tank. A total of 136 of these tanks were constructed and delivered to Kennedy Space Center, and 135 flew.

The ships were NASA-owned, and operated by United Space Alliance, a contractor. After the Shuttle was decommissioned, these ships were transferred to the Department of Transportation. They have two main engines of 2,900 horsepower combined, and could tow

60,000 pounds. They had added water jets that enhanced maneuverability in the Banana River, heading up to the Kennedy Spaceport. The recovery ships mission started at 24 hours before launch, when they left the Kennedy Launch site for the recovery zone. Each ship carried a crew of 10, with eight divers. Each ship was custom built for its job, in 1981.

The boosters were separated from the Shuttle at T plus 2 minutes, 7 seconds, having done their job of getting the vehicle up high and going fast. Parachutes deployed, slowing the 165,000 booster casings to 62 mph.

After rendezvous with the boosters, the parachutes are recovered and put on deck. A frustrum at the base of the booster is recovered. This is the section that housed the chutes. Now, 8 divers in two small boats go to work, installing a plug in the nozzle, and hooking up an air line to pump the water from the booster. After that is completed, the booster floats on the surface, and the ship can attach a line for towing,

The Shuttle vehicle itself was a winged rocket plane. It's three main engines burned liquid hydrogen and liquid oxygen form the large external tanks, which as later jettisoned into the ocean. Two solid rocket boosters provided a good start. The Shuttle had its own propulsion system to adjust orbit and attitude – the OMS, orbital maneuvering system. Upon re-entry, the Shuttle was a big, heavy glider. It has aerodynamic control surfaces, and returned to an airport.

Nominally, the Shuttle would return to the airport at Kennedy Space Center, where it could be readied for its next flight. An alternate site was at the dry lake beds at Edwards Air Force Base in California. This would be used if the KSC runway was not available, or there was a in-flight contingency that could require the larger landing field at Edwards. Then, the Shuttle would have to be hoisted on top to the 747 carrier aircraft for the flight back to Kennedy,

There were several contingency airfield around the globe addressing contingencies at launch. The first was, after the solid rocket boosters burned out, the main engines would be shut down, and the external tank jettisoned. The Shuttle would turn back and land the the Kennedy runway.

The next generation crewed capsule, the 4-person Orion, is designed to land in the ocean, like the Apollo capsule. The recovery method will involve a Navy amphibious landing ship, which can tow the spacecraft onboard. Testing is ongoing in the pacific, using the USS San Diego.

Shuttle External Tank transport
The Michoud facility is administratively a part of the Marshall Space Flight Center, and was built to support construction and integration of the massive Saturn-V moon rockets. These were so large they were delivered by barge to the Kennedy Space Center. For the Space

Shuttle Program, Michoud was used to assemble the large external fuel tank. It is located east of New Orleans, Louisiana. The barge was called the Pegasus, and it was towed by the same ships used for booster recovery in the Atlantic. A second winch was added for external tank towing. The ships need a 12 day round trip from Port Canaveral to the Michoud Assembly Facility and back. The tank is conveyed under cover on the barge *Pegasus,* which has its own crew of three.

Airborne Observatory

This section discusses platforms with a telescope, including planes, and airships. Infrared observing missions can fly above the blocking water vapor in the atmosphere. The first balloon borne astronomy mission was Stratoscope I in 1957. The Boomerang (Balloon Observations Of Millimetric Extragalactic Radiation And Geophysics) project looked at the cosmic background microwave radiation. It flew above 125,000 feet. First flight was in 1997. In 1998 and 2003, it ascended from McMurdo Station in Antarctica. The mission duration on this flight was 2 weeks.

NASA's Galileo Airborne Observatory was hosted on a Convair 990 aircraft. It was frst used in 1965, but was unfortunately destroyed n a mid-ari collision in 1973.

NASA's Kuiper Airborne Observatory used the large Lockheed C-141 Starlifter. It operated from 1974 to 1995.

The Joint NASA-DLR (Stratospheric Observatory for Infrared Astronomy (SOFIA) has been in use since 2010.

It uses a Boeing 747SP with a 2.7 meter telescope. The aircraft is based at the Armstrong Flight Research Center in Pasadena, CA. The telescope is specially mounted to isolated it from movements of the plane.

Wrap-up

The amount of ground-based infrastructure it took to support early missions back in the Apollo-era was staggering. It all had to be built from scratch, with no models, such as the Internet, to follow. A world-wide data infrastructure was required, and achieved. This required both aircraft and ships to fill in the coverage gaps. Now, it is all done from satellites in a higher orbit, and real-time video from the ISS is available on our smart phones. Most of the data travels on the ubiquitous Internet.

Bibliography

Anderson, John *X-15: The World's Fastest Rocket Plane and the Pilots Who Ushered in the Space Age*, Zenith Press, 2014, ISBN-0760344450.

The Astronauts, *We Seven*, 1962, Simon and Schuster, 2010, Renewal Edition, ISBN-1439181039.

Baker, David NASA *Mercury - 1956 to 1963 (all models)- An insight into the design and engineering of Project Mercury - America's first manned space programme*, 2017, Haynes Publishing, ISBN-1785210645.

Baker, David NASA *Space Shuttle Manual: An Insight into the Design, Construction and Operation of the NASA Space Shuttle*, Zenith Press, 2011, ISBN-0760340765.

Chambers, Mark A. *Flight Research at NASA Langley Research Center (VA)*, Images of Aviation, Arcadia Press, 2007, ISBN-073854437X.

Collins, William R. *Histories of the Space Tracking and Data Acquisition Network (STADAN), the Manned Space Flight Network (MSFN), and the NASA Communications Network (NASCOM)*, 1974, ASIN-B000736FAO.

Corliss, William R. *The evolution of the satellite tracking and data acquisition network STADAN (Goddard* historical note), 1967, ASIN-B0007I2QZC.

Corliss, William R. *The Space Tracking and Data Acquisition Network (STADAN), The Manned Space Flight Network (MSFN) and the NASA Communications Network (NASCOM)*, 1974, avail: https://ntrs.nasa.gov/archive/nasa/casi.ntrs.nasa.gov/19750002909.pdf

Corliss, William R. *Spacecraft Tracking*, 1969, ASIN-B002K7SK44.

Corliss, William R. *The Evolution of the Satellite Tracking and Data Acquisition Network (STADAN)*, 1967, NASA X-202-67-26

Greene, John A. To Fly *What Others Only Imagine: NASA Dryden Flight Research Center,* Historic Aircraft, 2012, ASIN-B008L41376.

Hallion, Richard P.; Gorn, Michael H. *On the Frontier: Experimental Flight at NASA Dryden*, 2003, Smithsonian, ISBN-1588341348.

Hansen, James R. *Engineer in Charge: A History of the Langley Aeronautical Laboratory, 1917-1958*, (NASA History Series Book 4305), 2013, ASIN-B00CJJIWHS.

Hansen, James R. *Spaceflight Revolution: NASA Langley Research Center from Sputnik to Apollo* (NASA History Series Book 4308), 2013, ASIN-B00BT4GO38.

Jenkins, Dennis R., Lands, Tony R. *Hypersonic: The Story of the North American X-15,* 2003, Specialty Press, ISBN-1580071317.

Kovalchik, Dan *Range Rats at Sea,* 2001, ASIN-B007N0PW0S.

Kovalchik, Dan *The Devil's Ashpit: and other Tales of Ascension Island,* 2014, ASIN-B00LDQ39ZS.

McKenna, John J. "NASA Tracking Ship Navigation Systems," 1975, avail: http://tycho.usno.navy.mil/ptti/1975papers/Vol%2007_35.pdf

Miller, Jay and Yeager, Chuck, *The X Planes: X-1 to X-31*, Aerofax Inc.; Revised edition, 1988, ISBN-0517567490.

NASA, *Venture into Space: Early Years of Goddard Space Flight Center - Report on Vanguard, Mercury Tracking, Explorer, Pioneer, Tiros, Telstar, Relay, and Syncom Satellites* (NASA SP-4301), 2017, ISBN-1973161672.

NASA/JSC, NASA Facts: *Tracking and Communications*, 1971, ASIN-B001K94HFO.

NASA, *Handbook for Apollo Instrumentation Ships,* 1968, NASA MG-402, avail:

https://www.hq.nasa.gov/alsj/MG-402-Ships-Manual-OCR.pdf

Reichl, David *Project Mercury: America in Space Series*, 1st Edition, Schiffer; 2016, ISBN-0764350692.

Reichl, David *Project Gemini,* America in Space Series, 1st. Edition, Schiffer, 2016, ISBN-0764350706.

Rumerman, Judy A., NASA *Historical Data Book: Volume VIII: NASA Earth Science and Space Applications, Aeronautics, Technology, and Exploration, Tracking and Data Resources 1989-1998* (The NASA History Series), 2014, ISBN-1501062018.

Spennemann, Dirk H,. R. *The Naval heritage of the US space programme: A case of losses,* J. Maritime Research, 2011, avail: https://doi.org/10.1080/21533369.2005.9668350.

Stakem, Patrick H. *Rocketplanes to Space,* 2017, PRRB Publishing, ISBN-1549992589.

Storms, Dr. Harrison A. and Hallion, Dr. Richard P. X-15: Reaching for Space (The X-Plane Series), 2013, ASIN-B00CLIH4Q2.

Thompson, Milton O.; Armstrong, Neil A. *At the Edge of Space: The X-15 Flight Program*, 2013, Smithsonian Books, ASIN-B00DFIDW9K.

Tsiao, Sunny, *Read You Loud and Clear: The Story of NASA's Spaceflight Tracking and Data Network* (NASA History), 2008, ISBN-978-0160801914.

Resources

- http://www.collectspace.com/ubb/Forum29/HTML/000721-2.html

- http://www.rangerat.com/

- http://www.collectspace.com/ubb/Forum29/HTML/000721-2.html

- T-AGM-19 / T-AG-194 Vanguard, https://fas.org/man/dod-101/sys/ship/t-agm-19.htm

- www.nasa.gov

- https://history.nasa.gov/ tindex.html#5

- https://www.nasa.gov/exploration/systems/orion/index.html

- https://history.nasa.gov/tindex.html#5

- http://klabs.org/history

- Encyclopedia Astronautica, http://www.astronautix.com/

- Vectors website - http://vc.airvectors.net/idx_sci.html

- NASA, Solid Rocket Booster Recovery, avail: https://spaceflight.nasa.gov/shuttle/support/processing/srb/

- Handbook for Apollo Instrumentation ships, avail, https://www.hq.nasa.gov/alsj/MG-402-Ships-Manual-OCR.pdf

- Navy Recovery Ships, avail: https://history.nasa.gov/ships.html

- Early spacecraft tracking ships, avail: http://www.collectspace.com/ubb/Forum29/HTML/000721-2.html

- Liberty Star and Freedom Star, Shuttle SRB recovery and ET towing.
  avail, https://www.nasa.gov/mission_pages/shuttle/behindscenes/recovery_ships.html

- Bendix Field Engineering, http://www.bfec.us/

- http://spacecovers.com/jpers/zjp_emp_rca_etr_rangerat_ships.htm

- https://www.sofia.usra.edu/

- wikipedia, various.

Directory of U. S. Navy ships used to recover NASA crewed missions.

From, http://tycho.usno.navy.mil/ptti/1975papers/Vol%2007_35.pdf

This list includes only vessels designated as part of the official Naval recovery force. Tracking ships are also not included in this list. PRS = primary recovery ship, an aircraft carrier.

Mercury-Redstone 3 (MR-3) - May 5, 1961
USS Lake Champlain (PRS)
USS Abbot
USS Ability
USS Decatur
USS Newman K. Perry
USS Notable
USS Recovery
USS Rooks
USS Sullivans
USS Wadleigh

Mercury-Redstone 4 (MR-4) - July 21, 1961
USS Randolph (PRS)
USS Alacrity
USS Conway
USS Cony
USS Exploit
USS Lowry

USS Recovery
USS Stormes

Mercury-Atlas 6 (MA-6) - February 20, 1962
USS Noa (actual recovery ship)
USS Randolph (PRS)
USS Antietam
USS Bearss
USS Barry
USS Blandy
USS Brownson
USS Chuckawan
USS Cone
USS Constellation
USS Exploit
USS Glennon
USS Goodrich
USS Hugh Purvis
USS Kenneth D. Bailey
USS Norfolk
USS Observer
USS Recovery
USS Sarsfield
USS Charles S. Sperry
USS Stormes
USS Stribling
USS Turner
USS Witek USS Antietam

Mercury-Atlas 7 (MA-7) - May 24, 1962
USS John R. Pierce (actual recovery ship)

USS Intrepid (PRS)
USS Barton
USS Dewey
USS Donner
USS Ellison
USS Elokomin
USS English
USS Farragut
USS Forrestal
USS Fred T. Berry
USS Hank
USS Hois
USS Hunt
USS Massey
USS Moale
USS Remey
USS Robinson
USS Shenandoah
USS Soley
USS Spiegel Grove
USS Sturdy
USS Swerve
USS Wren

Mercury-Atlas 8 (MA-8) - October 3, 1962
USS Kearsarge (PRS)
USS Affray
USS Alacrity
USS Barry
USS Bordelon
USS Charles Adams

USS Charles S. Sperry
USS Decatur
USS Dyess
USS Epperson
USS Fletcher
USS Fred T. Berry
USS Furse
USS Haynesworth
USS Henley
USS Hoist
USS Independence
USS Ingraham
USS John Paul Jones
USS Kaskaskia
USS Lake Champlain
USS Norris
USS O'Bannon
USS Phillip
USS Radford
USS Walker
USS Willard Keith

Mercury-Atlas 9 (MA-9) - May 15-16, 1963
USS Kearsarge (PRS)
USS Adroit
USS Beatty
USS Chipola
USS Compton
USS Davis
USS De Haven
USS Duncan

USS Epperson
USS Fletcher
USS Frank Knox
USS Gainard
USS Harwood
USS Hyman
USS John A. Bole
USS John W. Thomason
USS Kawishiwi
USS Lofberg
USS Mansfield
USS Myles C. Fox
USS Opportune
USS Stalwart
USS Taussig
USS Wasp

Gemini-Titan 3 (GT-3) - March 23, 1965
USS Intrepid (PRS)
USS Ault
USS Bigelow
USS Boston
USS Cony
USS Diligence
USS Douglas H. Fox
USS Harold J. Ellison
USS Harwood
USS John Paul Jones
USS Kankakee
USS Mullinix
USS Nipmuc

USS Rich
USS Robert L. Wilson
USS Robert Owens
USS Sarsfield
USS Sturdy
USS Swerve
USCGC Vigilant (US Coast Guard Cutter)

Gemini-Titan 4 (GT-4) -June 3-7, 1965
USS Wasp (PRS)
USS Barry
USS Blandy
USS Charles S. Sperry
USS Chukawan
USS Furse
USS Goldsborough
USS Hawkins
USS Higbee
USS Hoist
USS Leonard F. Mason
USS Nimble
USS Orleck
USS Ponchatoula
USS Rich
USS Robert A. Owens
USS Rupertus
USS Skill

Gemini-Titan 5 (GT-5) -August 21-29, 1965
USS Lake Champlain (PRS)
USS Avenge

USS Chipola
USS DuPont
USS Exultan
USS George MacKenzie
USS Goldsborough
USS James C. Owens
USS John W. Weeks
USS Leonard F. Mason
USS Manley
USS Neosho
USS New
USS Preserver
USS Taylor
USS Waldron

Gemini-Titan 6A (GT-6A) - December 15-16, 1965
USS Wasp (PRS)
USS Ability
USS Aucilla
USS Cochrane
USS George MacKenzie
USS Joseph P. Kennedy
USS Meredith
USS Paiute
USS Ponchatoula
USS Power
USS Renshaw
USS Rupertus
USS Waccamaw
USS Waldron

Gemini-Titan 7 (GT-7) - December 4-18, 1965
USS Wasp (PRS)
USS Ability
USS Aucilla
USS Cochrane
USS George MacKenzie
USS Joseph P. Kennedy
USS Meredith
USS Paiute
USS Ponchatoula
USS Power
USS Renshaw
USS Rupertus
USS Waccamaw
USS Waldron

Gemini-Titan 8 (GT-8) -March 16, 1966
USS Leonard F. Mason (actual recovery ship)
USS Boxer (PRS)
USS Caloosahatchee
USS Charles P. Cecil
USS Cochrane
USS Fidelity
USS George K. MacKenzie
USS Goodrich
USS Hassayampa
USS Myles C. Fox
USS Noa
USS Paiute

Gemini-Titan 9A (GT-9A) - June 3-6, 1966

USS Wasp (PRS)
USS Bordelon
USS Chikaskia
USS Epperson
USS George K. MacKenzie
USS Hassayampa
USS McCaffery
USS Nimble
USS Opportune
USS Papago
USS Robert L. Wilson
USS Rupertus
USS Sabine
USS Saratoga
USS William C. Lawe

Gemini-Titan 10 (GT-10) - July 18-21, 1966
USS Guadalcanal (PRS)
USS Allen M. Sumner
USS Benjamin Stoddert
USS Collet
USS DeHaven
USS Kawishiwi
USS Norris
USS Opportune
USS Severn
USS William C. Lawe

Gemini-Titan 11 (GT-11) - September 12-15, 1966
USS Guam (PRS)
USS Forest Royal

USS Kawishiwi
USS Mansfield
USS McCaffery
USS Nimpuc
USS O'Bannon
USS O'Brien
USS Severn
USS Theodore Chandler
USS Wallace L. Lind

Gemini-Titan 12 (GT-12) - November 11-15, 1966
USS Wasp (PRS)
USS Canisteo
USS Charles H. Roan
USS Hollister
USS Joseph P. Kennedy
USS Joseph Strauss
USS Kankakee
USS Kawishiwi
USS Lloyd Thomas
USS Ozbourn
USS Preserver

Apollo-Saturn 7 (AS-7) - October 11-22, 1968
USS Essex (PRS)
USS Arneb
USS Cambria
USS Cochrane
USS Henry Tucker
USS Nicholas
USS Paiute

USS Ponchatoula
USS Rupertus

Apollo-Saturn 8 (AS-8) - December 21-27, 1968
USS Yorktown (PRS)
USS Arlington
USS Chipola
USS Chuckawan
USS Cochrane
USS Francis Marion
USS Guadalcanal
USS Nicholas
USS Rankin
USS Rupertus
USS Salinan
USS Sandoval

Apollo-Saturn 9 (AS-9) - March 3-13, 1969
USS Guadalcanal (PRS)
USS Algol
USS Cochrane
USS Leonard F. Mason
USS Nicholas
USS Paiute

Apollo-Saturn 10 (AS-10) -May 18-26, 1969
USS Princeton (PRS)
USS Arlington
USS Carpenter
USS Chilton
USS Ozark

USS Rich
USS Salinan

Apollo-Saturn 11 (AS-11) -July 16-24, 1969
USS Hornet (PRS)
USS Arlington
USS Carpenter
USS Goldsborough
USS Hassayampa
USS New
USS Ozark
USS Salinan

Apollo-Saturn 12 (AS-12) - November 14-24, 1969
USS Hornet
USS Austin
USS Escape
USS Hawkins
USS Joseph J. Strauss

Apollo-Saturn 13 (AS-13) April 11-17, 1970
USS Iwo Jima (PRS)
USS Benjamin Stoddert
USS Bordelon
USS Escape
USS Forest Royal
USS Granville S. Hall
USS Kawishiwi
USS New
USS William C. Lawe

Apollo-Saturn 14 (AS-14) - January 31-February 9, 1971
USS New Orleans (PRS)
USS Carpenter
USS Hawkins
USS Paiute
USS Ponchatoula
USS Spiegel Grove

Apollo-Saturn 15 (AS-15) - July 26-August 7, 1971
USS Okinawa (PRS)
USS Austin
USS Kawishiwi
USS Salinan

Apollo-Saturn 16 (AS-16) - April 16-27, 1972
USS Ticonderoga (PRS)
USS Alacrity
USS Exploit
USS Goldsborough
USS Opportune
USS Ponchatoula

Apollo-Saturn 17 (AS-17) - December 7-19, 1972
USS Ticonderoga (PRS)
USS Camden
USS Recovery
USS Saginaw

Skylab-2 (SL-2) -May 25-June 22, 1973
USS Ticonderoga (PRS)
USS Escape

USS Grapple

Skylab-3 (SL-3) -July 28- September 25, 1973
USS New Orleans (PRS)
USS Escape
USS Grapple

Skylab-4 (SL-4) -November 16, 1973- February 8, 1974
USS New Orleans (PRS)
USS Opportune

Apollo-Soyuz Test Project (ASTP) July 15-24, 1975
USS New Orleans (PRS)

Glossary of Aerospace Terms

AAFB – Auxiliary Air Force Base

AF – (U.S.) Air Force

AFB – Air Force Base

AIAA – American Institute of Aeronautics and Astronautics.

AOMC – Army Ordnance Missile Command – 1958

AOS – acquisition of signal

Apogee – farthest point in the orbit from the Earth.

ARPA – Advanced Research projects Agency.

ASIN – Amazon Standard Inventory Number

ASN – Acquisition and Stabilization

Astrionics – electronics for space flight.

ATS-6 – Applications Technology Satellite – 6.

Bathymetric – refers to depth of ocean floor.

BNS – Bathymetric Navigation System

Boomerang - Balloon Observations Of Millimetric Extragalactic Radiation and Geophysics

CaDFISS – Computation and Data Flow Integrated Subsystem Test

C-band – 4 – 8 GHz.

CCSDS – Consultive Committee on Space Data Standards.

CEV - Crew Exploration Vehicle, part of Orion.

Cyrogenic – very low temperatures.

Dead-stick landing – landing without propulsion. Generally, "get it right the first time"

DLR - Deutsches Zentrum für Luft- und Raumfahrt German Aerospace Center.

DoD – (U. S.) Department of Defense.

DOT – dynamic operations test

DTM – dynamic test model, for structural tests.

ET – external tank (Shuttle)

ETR – Eastern Test Range, Cape Canerval.

EVA – Extra-vehicular Activity

GHz = giga (10^9) Hz.

Gimbal – pivoted support, allowing rotation about 1 axis.

Gpm – gallons per minute.

GSFC – NASA Goddard Space Flight Center, Greenbelt, MD.

Gyro – device to measure angular rate.

HRDS – High data rate system, part of NISN.

ICBM – Intercontinental Ballistic Missile.

INS – integrated navigation system.

IP – Internet protocol

IPOnet – Internet Protocol operational network, part of STDN.

IRBM – Intermediate Range Ballistic Missile.

IRIG – Inter-Range Instrumentation Group – a standards body.

ISBN – International Standard Book Number

ISP – specific impulse. Measure of efficiency of rocket engine. Units of seconds.

IST – integrated system test.

JPL – Jet Propulsion Laboratory, Pasadena, CA.

JSC – Johnson Space Center, Houston, Texas.

Jupiter – ICBM, 3-stage. Developed by von Braun Team.

KAO - Kuiper Airborne Observatory

Karman line – 100 km above the Earth's surface. The official definition of "space."

Kev – kilo electron volts, measure of energy of a particle.

KSC – NASA Kennedy Space Center, launch site,

Florida.

Lbf – pounds, force.

LC-37 – Launch Complex – 37 at KSC.

LEM – lunar excursion module.

LEO – low Earth orbit.

LES – Apollo Launch Escape System.

Loran-C – navigation system (pre-GPS) using land based transmitters.

LOS – loss of signal

Lox – liquid oxygen, boils at -297 F.

LRO – Lunar Reconnaissance Orbit. Ongoing lunar mapping and imaging mission.

LVDC – Launch Vehicle Digital Computer, Saturn

Mev – million electron volts, measure of energy of a particle.

MINITRACK – "Minimum Trackable Satellite " U. S. satellite tracking network, 1957.

M&O – manitenance & operations

MOC – Mission operations center.

MPCV – Multi-Purpose Crew Vehicle (NASA's Orion).

Mph – miles per hour.

MSFN – Manned Space Flight Network

MSC – Manned Space Center, Houston, TX. Renamed Johnson Space Center.

MSFC – NASA Marshall Space Flight Center, Huntsville, AL.

m/s – meters per second.

NACA – National Advisory Committee for Aeronautics, a precursor to NASA.

NASA – National Aeronautics and Space Administration.

NCC – Network Control Center (at GSFC).

NISN – NASA Integrated Services Network.

NORAD – North American Air Defense (Command) USAF.

OCC – operations control center.

OE – operator error.

Ops - operations

OV – orbiting vehicle

Perigee – closest point in the orbit from the Earth.

POCC – Payload operations control center.

PRS - primary recovery ship.

R&D – research & development.

Redstone – Army missile developed by the von Braun team. Used for Mercury manned flights.

Redstone Arsenal – Army R&D facility in Huntsville, AL. Later NASA-MSFC.

RP-1 – rocket propellant one, highly refined kerosene.

SAGE – Semi-Automatic Ground Equipment – early USAF computerized command & control

SATAN – Satellite Auto-tracking antenna

S-band – 2-4 GHz.

SCA – Shuttle carrier aircraft.

SIP – Ship's Instrumentation Manager.

SLS – Space Launch System – NASA next generation launch vehicle.

Sounding rocket – a smaller vehicle, usually sub-orbital, to take "soundings" of the atmosphere.

SPAMS – ship's position and attitude measurement system

SRB – (Shuttle) solid rocket booster.

SOFIA – Stratospheric Observatory for Infrared Astronomy.

SOM – Ship's operation s Manager.

SRT – Station readiness test.

SST – sub-system test.

ST – system test

STADAN – Space Tracking and Data Acquisition Network.

STDN – Space Flight Tracking and Data Network.

Stratospheric Observatory for Infrared Astronomy.

STS – Space Transportation System (Space Shuttle)

TDRS – Tracking and Data Relay Satellite

TDRSS – Tracking and Data Relay Satellite System

Titan – ICBM and NASA/USAF launch vehicle.

TM – Technical Manual.

USAF – United States Air Force.

USNS – U.S. Naval Ship

UT – Universal time.

VHF – very high frequency (MHz)

WAN – wide area network

WSMR – White Sands Missile Range, New Mexico. U. S. Army Facility.

ZOE – zone of exclusion.

Glossary of Nautical Terms and Phrases

from MG-402, Handbook for Apollo Instrumentation Ships.

ABAFT - behind, or farther aft. The mainmast is abaft the foremast.

ABEAM - at right angle s to the center line of the ship. Refers to an object outside the ship.

ABOARD - on or in a ship.

ACCOMMODATION LADDER - steps leading down a ship's side ; used for boarding.

ACKNOW LE DGEMENT - a statement that a message has been received.

AFT: at, near, or in the direction of the stern.

ALL HANDS - refers to every man (person) aboard .

AMIDSHIPS - in the line of the keel;sometimes halfway between the bow and stern.

ASTERN - behind the ship ; on a bearing of 1800 from ahead.

AUXILIARY - an assisting machine or vessel, such as an air-conditioning machine or a fuel ship.

BALLAST - heavy weights in the hold of a vessel or ship to increase stability by lowering center of gravity.

BEACON - an aid to navigation placed on or near a danger spot.

BEAM: the greatest width of a ship.

BEARING - the direction of an object expressed either in degrees either as relative or true bearing.

BELOW - beneath the main deck.

BERTH - a space for a ship to moor or anchor.

BILGE - lower part of a vessel where waste and seepage collect.

BILLET - an allotted place to sleep ; refers also to a particular man's duties aboard ship.

BLINKER - a set of lights at the masthead or on the end of a yardarm, connected to a telegraph key and used for sending flashing light signals .

BOAT: a small vessel which can be hoisted onto or carried by a ship.

BO'SN'S CHAIR - a seat, consisting of a short board fastened in the end of a line, on which a man may be suspended for working aloft or over the side.

BO'SN'S PIPE: a small , shrill whistle used by the bo'sn's mate.

BOW - forward part of a vessel .

BRIDGE - the raised pl platform in the forward part of the ship from which the ship is steered or navigated.

BULKHEAD - a partition separating compartments ; corresponds to a wall in a building.

BUOY - a floating marker moored to the bottom which by shape and color conveys navigational information.

CABIN - the captain's quarters .

CAPSIZE - to overturn in a small boat.

CAST OFF - to let go .

CENTER LINE - imaginary straight line running from the bow to the stern of a ship.

CHART - a nautical map used as an aid in navigation.

CLEAR - to leave a port with all formalities concluded; to empty; to work clear, as of a shoal; to untangle.

CLOSE ABOARD - near to the ship.
COLORS - the national ensign.
COMMAND - term applied to a ship or ships under one officer; a directive indicating what to do and how to do it.
COMPANION WAY - passageway on board ship.
COMPARTMENT - a space below deck between bulkheads, corresponding to a room of a building.
COUNTRY - the space near to a compartment or quarters, such as wardroom country or officers' country.
COURSE - the direction steered by a vessel expressed in degrees.
COXSWAIN - the enlisted man in charge of a boat and usually serving as steersman.
DAMAGE CONTROL - maintenance of watertight integrity of the ship during
battle or storms, including necessary repairs.
DAVIT - a curved metal spar fitting in a socket on the deck and projecting over the side of the ship for hoisting boats or handling weights.
DEAD AHEAD - directly ahead.
DEAD RECKONING - a navigator's estimate of the ship's position dependent upon course steered and distance run, independent of sights or bearings ; derived from "deduced reckoning."
DECK - corresponds to the floor of a building.
DOCK - a landing pier for boats or ships.
DOG - a type of bolt and nut used to secure watertight doors, hatch covers, or manhole covers.
DOLDDRUMS - belts on each side of the Equator in

which, ordinarily, little or no wind blows.

DOUBLE BOTTOMS - watertight subdivisions of a ship next to the keel and between the outer and inner bottoms.

DOWSE - to put out a light ; to cover with water.

DRAFT - the depth of water from the surface to the ship's keel ; a detail of men.

EBB TIDE - condition along the coast when the tide is going out.

EMBARK - to go on board ship.

EMERGENCY SPEED - all the speed of which a ship is capable.

ENSIGN - the national flag; a junior commissioned officer in the Navy.

EVEN KEEL - floating level.

FAIR TIDE - a tide running in the same direction as the ship.

FAIRWAY - an open channel.

FANTAIL - the part of the stern of the ship.

FAST - secure.

FATHOM - a unit of measurement equaling 6 feet.

FLANK SPEED - a certain prescribed increment above standard speed.

FLYING BRIDGE - a bridge on a ship having no supports and extending out from the control tower.

FORE AND AFT - running in the direction of the keel .

FORECASTLE - the upper deck forward of the foremast; a forward compartment where the crew lives.

FRAME - the ribs of a ship strengthening and supporting the plating.

FUNNEL - the smokestack of a ship.

GALLEY - the ship's kitchen.
GANGPLANK - a portable bridge used for boarding a ship from a dock
GIG - ship's boat used by the commanding officer.
GUNWHALE - the upper edge or rail of a ship or boat's side.
HAND - a member of the crew.
HANDY -BILLY - a small portable pump.
HATCH - an opening in the ship's deck
HEAD - a ship's toilet.
HELM - a tiller
HIGH SEAS - the entire ocean beyond the three-mile limit where no nation has military authority
LINE OFFICER – an officer who holds military authority in the chain of command.
HULL - framework of a vessel together
MAIN DECK - the highest complete deck
INBOARD: toward the center line of the extending from stem to stern and side to side.
ISLAND - superstructure containing conning tower, navigation bridge, communication platform, flying bridge, etc.
JETSOM - goods which sink when thrown overboard at sea.
JETTY - a landing wharf or pier, a dike at a river's mouth.
KEEL - the backbone of a ship running from stem to sternpost at the bottom.
KNOT - one nautical mile per hour.
LADDER - a metal, wooden, or rope stairway.
LAND FALL - first sighting of land at the end of a sea

voyage.

LASH - to tie or secure.

LEAVE - special permission to be absent from ship or station.

LEE - away from the direction of the wind.

LIFELINE - a line secured along the deck to lay hold of in heavy weather;

MOORING - securing a ship to a dock or buoy, or anchoring with two anchors.

NAUTICAL MILE - the length of a minute of a great circle of the earth or 6,080.20 feet.

OFFICER OF THE DECK - the officer in charge of the ship during each watch and on deck as the captain's representative.

OVERHEAD - equivalent to the ceiling in a building.

PASSAGEWAY - a corridor or hallway onboard ship.

PILOT - an expert who conducts ships in and out of harbors and in dangerous waters.

POOP DECK - the partial raised deck and after structure at the stern over the main deck of a vessel.

PORT - the left side of a ship facing forward

QUARTERDECK - the part of the upper

STARBOARD - the right side of a ship deck reserved for honors and cere- looking forward.

STATION BILL - a bill listing stations of the crew at emergency drills.

QUARTERS - living space ; all hands assembled at established stations for review.

REEF - a chain or ridge of racks, coral, or sand in shallow water.

RUNNING LIGHTS - lights required by law which are

carried by a ship underway.
SCREW - propeller.
SCUTTLEBUTT - a container of water for drinking purposes; a rumor.
SEA KEEPING - the ability of a ship to stay at sea for long periods of time.
SECOND DECK - a complete deck next below the m main deck.
SECURE - to m make fast; to tie ; order given on completion of a drill or exercise on board ship.
SHAKEDOWN CRUISE - cruise of a newly commissioned ship to test all machinery and train the crew.
SICKBAY - ship 's hospital or dispensary
SKIPPER - slang for the captain.
SMOKING LAMP - the expression, "the smoking lamp it lit," means permission is given to smoke.
SOUND-POWERED PHONE - a phone powered by voice ; standard battle phone.
SQUARE A WAY - to get things settled down or in order ; to complete a job .
Staff officer – an officer whose duties are special rather than military; doctor or chaplain.
Starboard – right side of the ship, looking forward.
Station bill –a bill listing stations of the crew at emergency drills
Stern – after part of the ship
Superstructure – equipment and fittings, except armament, extending above the hull.
Swab – a rope mop
Swell – heave of the sea.

TAFFRAIL - a rail at the stern of a ship .

TOPSIDE - above decks.

UNDER WAY - a vessel is said to be underway when she is not at anchor nor made fast to the shore, nor aground.

UNDERTOW - a seaward current near the bottom in heavy surf.

WAKE - the track or trail in water which a vessel leaves behind.

WARDROOM - officers' assembly and mess room aboard a ship.

WATCH - a post or period of duty.

WATCH, QUARTER, AND STATION BILL - a list or chart giving the watch duties, billet and emergency, or battle station of every man aboard ship.

WEATHER DECK - the portion of the main, forecastle, poop, and upper decks which is exposed to the elements.

WINDWARD - into the wind ; toward the direction from which the wind is blowing.

YARD - shipbuilding and repair depot for Navy vessels- as Boston Navy Yard .

YAW - to steer badly, zigzagging back and forth across the intended course.

If you enjoyed this book, you might find something else from the author interesting as well.

- Stakem, Patrick H. *16-bit Microprocessors, History and Architecture*, 2013 PRRB Publishing, ISBN-1520210922.

- Stakem, Patrick H. *4- and 8-bit Microprocessors, Architecture and History*, 2013, PRRB Publishing, ISBN-152021572X,

- Stakem, Patrick H. *Apollo's Computers,* 2014, PRRB Publishing, ISBN-1520215800.

- Stakem, Patrick H. *The Architecture and Applications of the ARM Microprocessors,* 2013, PRRB Publishing, ISBN-1520215843.

- Stakem, Patrick H. *Earth Rovers: for Exploration and Environmental Monitoring,* 2014, PRRB Publishing, ISBN-152021586X.

- Stakem, Patrick H. *Embedded Computer Systems, Volume 1, Introduction and Architecture*, 2013, PRRB Publishing, ISBN-1520215959.

- Stakem, Patrick H. *The History of Spacecraft Computers from the V-2 to the Space Station*, 2013, PRRB Publishing, ISBN-1520216181.

- Stakem, Patrick H. *Floating Point Computation*, 2013, PRRB Publishing, ISBN-152021619X.

- Stakem, Patrick H. *Architecture of Massively Parallel Microprocessor Systems*, 2011, PRRB Publishing, ISBN-1520250061.

- Stakem, Patrick H. *Multicore Computer Architecture,* 2014, PRRB Publishing, ISBN-1520241372.

- Stakem, Patrick H. *Personal Robots*, 2014, PRRB Publishing, ISBN-1520216254.

- Stakem, Patrick H. *RISC Microprocessors, History and Overview,* 2013, PRRB Publishing, ISBN-1520216289.

- Stakem, Patrick H. *Robots and Telerobots in Space Application*s, 2011, PRRB Publishing, ISBN-1520210361.

- Stakem, Patrick H. *The Saturn Rocket and the Pegasus Missions, 1965,* 2013, PRRB Publishing, ISBN-1520209916.

- Stakem, Patrick H. *Visiting the NASA Centers, and Locations of Historic Rockets & Spacecraft,* 2017, PRRB Publishing, ISBN-1549651205.
- akem, Patrick H. *Microprocessors in Space*, 2011, PRRB Publishing, ISBN-1520216343.

- Stakem, Patrick H. Computer *Virtualization and the Cloud*, 2013, PRRB Publishing, ISBN-152021636X.

- Stakem, Patrick H. *What's the Worst That Could Happen? Bad Assumptions, Ignorance, Failures and Screw-ups in Engineering Projects*, 2014, PRRB Publishing, ISBN-1520207166.

- Stakem, Patrick H. *Computer Architecture & Programming of the Intel x86 Family*, 2013, PRRB Publishing, ISBN-1520263724.

- Stakem, Patrick H. *The Hardware and Software Architecture of the Transputer*, 2011, PRRB Publishing, ISBN-152020681X.

- Stakem, Patrick H. *Mainframes, Computing on Big Iron*, 2015, PRRB Publishing, ISBN- 1520216459.

- Stakem, Patrick H. *Spacecraft Control Centers*, 2015, PRRB Publishing, ISBN-1520200617.

- Stakem, Patrick H. *Embedded in Space*, 2015, PRRB Publishing, ISBN-1520215916.

- Stakem, Patrick H. *A Practitioner's Guide to RISC Microprocessor Architecture*, Wiley-Interscience, 1996, ISBN-0471130184.

- Stakem, Patrick H. *Cubesat Engineering*, PRRB Publishing, 2017, ISBN-1520754019.

- Stakem, Patrick H. *Cubesat Operations*, PRRB Publishing, 2017, ISBN-152076717X.

- Stakem, Patrick H. *Interplanetary Cubesats*, PRRB Publishing, 2017, ISBN-1520766173 .

- *Stakem, Patrick H. Cubesat Constellations, Clusters, and Swarms, Stakem,* PRRB Publishing, 2017, ISBN-1520767544.

- Stakem, Patrick H. *Graphics Processing Units, an overview*, 2017, PRRB Publishing, ISBN-1520879695.

- Stakem, Patrick H. *Intel Embedded and the Arduino-101, 2017,* PRRB Publishing, ISBN-1520879296.

- Stakem, Patrick H. *Orbital Debris, the problem and the mitigation*, 2018, PRRB Publishing, ISBN-*1980466483*.

- Stakem, Patrick H. *Manufacturing in Space*, 2018, PRRB Publishing, ISBN-1977076041.

- Stakem, Patrick H. *NASA's Ships and Planes*, 2018, PRRB Publishing, ISBN-1977076823.

- Stakem, Patrick H. *Space Tourism*, 2018, PRRB Publishing, ISBN-1977073506.

- Stakem, Patrick H. *STEM – Data Storage and Communications*, 2018, PRRB Publishing, ISBN-1977073115.

- Stakem, Patrick H. *In-Space Robotic Repair and Servicing*, 2018, PRRB Publishing, ISBN-1980478236.

- Stakem, Patrick H. *Introducing Weather in the pre-K to 12 Curricula, A Resource Guide for Educators*, 2017, PRRB Publishing, ISBN-1980638241.

- Stakem, Patrick H. *Introducing Astronomy in the pre-K to 12 Curricula, A Resource Guide for Educators*, 2017, PRRB Publishing, ISBN-198104065X.
- Also available in a Brazilian Portuguese edition, ISBN-1983106127.

- Stakem, Patrick H. *Deep Space Gateways, the Moon and Beyond*, 2017, PRRB Publishing, ISBN-1973465701.

- Stakem, Patrick H. *Exploration of the Gas Giants, Space Missions to Jupiter, Saturn, Uranus, and Neptune*, PRRB Publishing, 2018, ISBN-9781717814500.

- Stakem, Patrick H. *Crewed Spacecraft*, 2017, PRRB Publishing, ISBN-1549992406.

- Stakem, Patrick H. *Rocketplanes to Space*, 2017, PRRB Publishing, ISBN-1549992589.

- Stakem, Patrick H. *Crewed Space Stations,* 2017, PRRB Publishing, ISBN-1549992228.

- Stakem, Patrick H. *Enviro-bots for STEM: Using Robotics in the pre-K to 12 Curricula, A Resource Guide for Educators,* 2017, PRRB Publishing, ISBN-1549656619.

- Stakem, Patrick H. *STEM-Sat, Using Cubesats in the pre-K to 12 Curricula, A Resource Guide for Educators*, 2017, ISBN-1549656376.

- Stakem, Patrick H. *Lunar Orbital Platform-Gateway,* 2018, PRRB Publishing, ISBN-1980498628.

- Stakem, Patrick H. *Embedded GPU's*, 2018, PRRB Publishing, ISBN- 1980476497.

- Stakem, Patrick H. *Mobile Cloud Robotics*, 2018, PRRB Publishing, ISBN- 1980488088.

- Stakem, Patrick H. *Extreme Environment Embedded Systems,* 2017, PRRB Publishing,

ISBN-1520215967.

- Stakem, Patrick H. *What's the Worst, Volume-2*, 2018, ISBN-1981005579.

- Stakem, Patrick H., *Spaceports*, 2018, ISBN-1981022287.

- Stakem, Patrick H., *Space Launch Vehicles*, 2018, ISBN-1983071773.

- Stakem, Patrick H. *Mars*, 2018, ISBN-1983116902.

- Stakem, Patrick H. *X-86, 40th Anniversary ed*, 2018, ISBN-1983189405.

- Stakem, Patrick H. *Lunar Orbital Platform-Gateway*, 2018, PRRB Publishing, ISBN-1980498628.

- Stakem, Patrick H. *Space Weather*, 2018, ISBN-1723904023.

- Stakem, Patrick H. *STEM-Engineering Process*, 2017, ISBN-1983196517.

- Stakem, Patrick H. *Space Telescopes,* 2018, PRRB Publishing, ISBN-1728728568.

- Stakem, Patrick H. *Exoplanets*, 2018, PRRB

Publishing, ISBN-9781731385055.

- Stakem, Patrick H. *Planetary Defense*, 2018, PRRB Publishing, ISBN-9781731001207.

- Stakem, Patrick H. *Exploration of the Asteroid Belt*, 2018, PRRB Publishing, ISBN-1731049846.

- Stakem, Patrick H. *Terraforming*, 2018, PRRB Publishing, ISBN-1790308100.

- Stakem, Patrick H. *Martian Railroad,* 2019, PRRB Publishing, ISBN-1794488243.

- Stakem, Patrick H. *Exoplanets,* 2019, PRRB Publishing, ISBN-1731385056.

- Stakem, Patrick H. *Exploiting the Moon,* 2019, PRRB Publishing, ISBN-1091057850.

- Stakem, Patrick H. *RISC-V, an Open Source Solution for Space Flight Computers,* 2019, PRRB Publishing, ISBN-1796434388.

- Stakem, Patrick H. *Arm in Space*, 2019, PRRB Publishing, ISBN-9781099789137.

- Stakem, Patrick H. *Extraterrestrial Life*, 2019, PRRB Publishing, ISBN-978-1072072188.

- Stakem, Patrick H. *Space Command*, 2019,

PRRB Publishing, ISBN-978-1693005398.

- CubeRovers, A Synergy of Technologys, 2020, PRRB Publishing, ISBN-979-8651773138.

- Robotic Exploration of the Icy moons of the Gas Giants. 2020, PRRB Publishing, ISBN- 979-8621431006

- Hacking Cubesats, 2020, PRRB Publishing, ISBN-979-8623458964.

- History & Future of Cubesats, PRRB Publishing, ISBN-979-8649179386.

- Hacking Cubesats, Cybersecurity in Space, 2020, PRRB Publishing, ISBN-979-8623458964.

- Powerships, Powerbarges, Floating Wind Farms: electricity when and where you need it, 2021, PRRB Publishing, ISBN-979-8716199477.

- Hospital Ships, Trains, and Aircraft, 2020, PRRB Publishing, ISBN-979-8642944349.

- *CubeRovers, a Synergy of Technologys*, 2020, ISBN-979-8651773138

- *Exploration of Lunar & Martian Lava Tubes by Cube-X*, ISBN-979-8621435325.

- *Robotic Exploration of the Icy moons of the Gas Giants*, ISBN- 979-8621431006.

- *History & Future of Cubesats*, ISBN-978-1986536356.

- *Robotic Exploration of the Icy Moons of the Ice Giants, by Swarms of Cubesats,* ISBN-979-8621431006.

- *Swarm Robotics,* ISBN-979-8534505948.

- *Introduction to Electric Power Systems*, ISBN-979-8519208727.

- *Centros de Control: Operaciones en Satélites del Estándar CubeSat* (Spanish Edition), 2021, ISBN-979-8510113068.

- *Exploration of Venus*, 2022, ISBN-979-8484416110.

- Patrick H. Stakem, *The Search for Extraterrestial Life,* 2019, PRRB Publishing, ISBN-1072072181.

- *The Artemis Missions, Return to the Moon, and on to Mars,* 2021, ISBN-979-8490532361.

- *James Webb Space Telescope. A New Era in Astronomy, 2021,* ISBN-979-8773857969.